CORAL SNAKES

DANGEROUS SNAKES

Tracy Nelson Maurer

TABLE OF CONTENTS

A Crabtree Seedlings Book

CRABTREE
Publishing Company
www.crabtreebooks.com

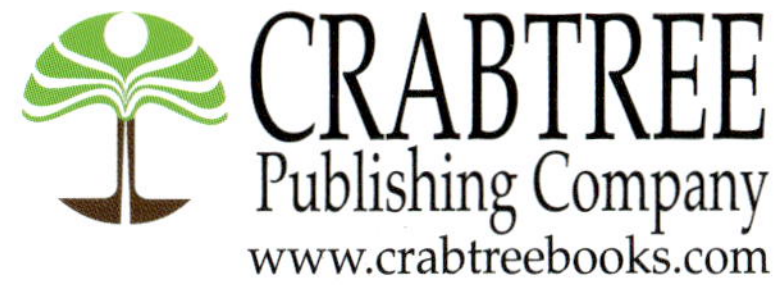

School-to-Home Support for Caregivers and Teachers

This book helps children grow by letting them practice reading. Here are a few guiding questions to help the reader with building his or her comprehension skills. Possible answers appear here in red.

Before Reading:

- What do I think this book is about?
    - *I think this book is about coral snakes.*
    - *I think this book is about colorful snakes.*

- What do I want to learn about this topic?
    - *I want to learn why coral snakes are dangerous.*
    - *I want to learn where coral snakes live.*

During Reading:

- I wonder why...
    - *I wonder why coral snakes hatch from eggs.*
    - *I wonder why coral snakes are small, with deadly venom.*

- What have I learned so far?
    - *I have learned that most coral snakes measure about 20 inches long.*
    - *I have learned that coral snakes are red, yellow, and black.*

After Reading:

- What details did I learn about this topic?
    - *I learned that coral snakes' fangs shoot venom into their prey.*
    - *I learned that they eat small lizards and other snakes.*

- Read the book again and look for the vocabulary words.
    - *I see the word* ***reptiles*** *on page 2 and the word* ***fangs*** *on page 12. The other glossary words are on pages 22 and 23.*

CORAL SNAKES

Coral snakes are small **reptiles**.

Baby coral snakes hatch from eggs.

When grown, most coral snakes measure about 20 inches (50 cm) long.

Some coral snakes are only as thick as a pencil.

Coral snakes' **scales** can be many colors.

Many coral snakes are known for their red, yellow, and black bands.

Red and yellow bands mean "Keep away!"

Milk snakes look similar to coral snakes. But unlike coral snakes, milk snakes do not have deadly **venom**.

Coral snakes have shorter **fangs** than most other snakes.

Fangs shoot the venom into their **prey**.

Coral snake venom is the most dangerous of all snake venom in the United States. The good news? Few coral snakes ever bite people.

Coral snakes hunt at night.

They eat small lizards and other snakes.

lizard

They rest under leaves and rocks or tuck into **burrows** and logs.

Glossary

burrows (BUR-ohz): Burrows are tunnels or holes in the ground.

fangs (FANGZ): Fangs are sharp, long teeth.

prey (PRAY): Prey is an animal that is hunted by another animal for food.

reptiles (REP-tilez): Reptiles are cold-blooded, scaly animals that breathe air.

scales (SKAYLZ): Scales are small pieces of hard skin that cover the bodies of reptiles and fish.

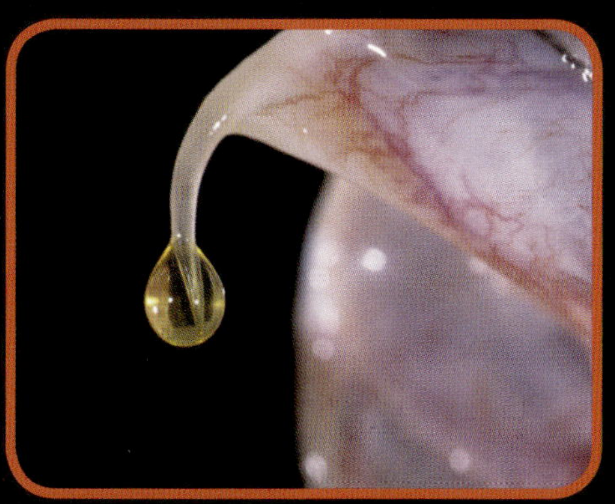

venom (VEN-uhm): Venom is a poison passed through a bite or sting.

Index

About the Author

Tracy Nelson Maurer

Tracy Nelson Maurer has written more than 100 books for young readers. She lives in Minnesota where it's too cold for most dangerous snakes.

Websites

https://www.nationalgeographic.com/animals/reptiles/e/eastern-coral-snake/
https://animals.net/coral-snake/

Written by: Tracy Nelson Maurer
Designed by: Jennifer Dydyk
Edited by: Kelli Hicks

Photographs:
mask for snakeskin graphic on cover and pages © shutterstock.com/Merydolla; yellow triangle with snake graphic © Top Vector Studio/Shutterstock; Cover photo: © shutterstock.com/ Jay Ondreicka; Page 3: ©shutterstock.com/Mark_Kostich; page 5 © Shutterstock.com/NOPPHARAT6395; page 7 © istock/ mspoli; page 9 © Luis Tejo | Dreamstime.com; page 10 (top) Shutterstock.com/vinap, (bottom) © istock/vinap; page 11 © Shutterstock/Matt Jeppson; page 13 Jason Ondreicka | Dreamstime.com; page 15 © Shutterstock/J.A. Dunbar; page 17 © Shutterstock/Jay Ondreicka; page 18 © Shutterstock/Kuznetsov Alexey; page 19 © Shutterstock/Martina Birnbaum; page 21 © Shutterstock/Luis César Tejo; page 22 (top) © Shutterstock/A_Lesik, (middle) Akarat Duangkhong | Dreamstime.com, (bottom) © Shutterstock/Karel Bartik; page 23 (middle) © istock/cturtletrax, (bottom) © Shutterstock/Joe McDonald

Library and Archives Canada Cataloguing in Publication

Title: Coral snakes / Tracy Nelson Maurer.
Names: Maurer, Tracy Nelson, 1965- author.
Description: Series statement: Dangerous snakes | "A Crabtree seedlings book". | Includes index.
Identifiers: Canadiana (print) 20210202246 | Canadiana (ebook) 20210202254 | ISBN 9781427162359 (hardcover) | ISBN 9781427159151 (softcover) | ISBN 9781427159168 (HTML) | ISBN 9781427159175 (EPUB) | ISBN 9781427159182 (read-along ebook)
Subjects: LCSH: Coral snakes—Juvenile literature.
Classification: LCC QL666.O64 M38 2022 | DDC j597.96/44—dc23

Library of Congress Cataloging-in-Publication Data

Names: Maurer, Tracy Nelson, 1965- author.
Title: Coral snakes / Tracy Nelson Maurer.
Description: New York : Crabtree Publishing, [2022] | Series: Dangerous snakes- a Crabtree seedlings book | Includes index.
Identifiers: LCCN 2021018863 (print) | LCCN 2021018864 (ebook) | ISBN 9781427162359 (hardcover) | ISBN 9781427159151 (paperback) | ISBN 9781427159168 (ebook) | ISBN 9781427159175 (epub) | ISBN 9781427159182
Subjects: LCSH: Coral snakes--Juvenile literature.
Classification: LCC QL666.O64 M39 2022 (print) | LCC QL666.O64 (ebook) | DDC 597.96/44--dc23
LC record available at https://lccn.loc.gov/2021018863
LC ebook record available at https://lccn.loc.gov/2021018864

Crabtree Publishing Company

www.crabtreebooks.com 1-800-387-7650

Printed in the U.S.A./062021/CG20210401

In Canada: We acknowledge the financial support of the Government of Canada through the Canada Book Fund for our publishing activities.

Published in the United States
Crabtree Publishing
347 Fifth Avenue, Suite 1402-145
New York, NY, 10016

Published in Canada
Crabtree Publishing
616 Welland Ave.
St. Catharines, Ontario L2M 5V6